The Infinite Number

The Infinite Number

Q. Apelles

Introduction

The task of the philosopher is to pursue as complete an understanding as human reason permits in a given subject. As it is generally acknowledged that a finite intellect can only reach so far in attempting to comprehend infinity before it must either abandon the effort or risk its own sanity, I have sought here to discover where the limits of this comprehension truly lie.

The primary aim of this thesis is to show that Galileo's insight into the nature of infinity, which I have attempted to develop and further explicate in this work, is fundamentally accurate. There does exist an actual infinite, and it is found in unity.

I present the investigation in the form of a Galilean dialogue in hope that the reader will follow the direction of my thoughts on the matter and adopt these speculations as his or her own, seeking from them not so much a conclusion as an understanding. The character of Salviati represents my own position throughout the course of the discussion, as he did for Galileo during the deliberations of the *First Day*. To Sagredo, the curious and intelligent layman, has been added the service of recalling to conversation important points and ideas from Galileo's own Two New Sciences. Simplicio has been replaced by the brighter and more mathematically inclined Literasto.

The setting of the dialogue takes place within the home of Salviati following the events of the *First Day*.

Evening of the First Day

Sagredo. You will think me overly ambitious, Salviati, for taking such advantage of the service you offered to Simplicio and me earlier this day, seeking you out now even in the quiet of your own home, where you are no doubt preparing to retire for the evening. To magnify the offense, I see that I am also intruding on company that you already entertain, and it is beyond me to ask both of you to suspend the academic discussion that is certainly taking place in order to satisfy my own intellectual desires.

Salviati. Your presence is always most welcome, Sagredo, no less now than before. For in my opinion, provided that we are not in too great a need of sleep, the tranquility of the night is every bit as hospitable to our cause as are the daylight hours, and, although I do have a guest, we truly had not a company until your own arrival. But tell me now what reflections bring you here, since I perceive this to be a matter of considerable importance.

Sagr. Although I still intend to allow my wits some recuperation from their daytime labors, our earlier conversation has left me with many things to turn over in my head. At the forefront of these is that mysterious notion of infinity, concerning which my mind will not rest without some further clarification.

In our inquiry into the nature of cohesion in solid bodies you presented us with demonstrations that offered many strange conclusions concerning the nature of the infinite, conclusions that granted you the liberty of positing an infinite number of infinitesimal voids as the cause of that cohesion.[1] Most astonishing, however, was your assertion that "to the extent that we go to greater numbers, by that much and more do we depart from the infinite number,"[2] and that the infinite in numbers is actually found in the number one.[3] I regret being unable to interject at that time, held captive as I was by the mysterious splendor of this conclusion. But since that initial awe has now subsided I feel it necessary to revisit the matter, to determine whether or not a relationship between unity

---

[1] Galileo Galilei, *Two New Sciences, Including Centers of Gravity and Force of Percussion*, trans. Stillman Drake (Toronto: Wall & Emerson, 1989), 26.
[2] Ibid., 45.
[3] Ibid.

and infinity truly exists, and to understand the consequences that would follow in recognition of such a connection. This is the task for which I now humbly request your assistance.

Salv. For my part, I cannot refuse the opportunity to return to the investigation of a subject so sublime, and whatever weariness may have begun creeping into my body has been banished by its presence. Moreover, I do not think that our enthusiasm will be lost on my visitor, a scholar of no ill repute and a man whose desire for knowledge is as insatiable as our own. Literasto is an old friend and a great benefactor of mine from years past.

Literasto. You will find me most willing to postpone our former discussion in order to investigate this new issue, since the chance to pursue such a wonderful topic in such devoted company does not occasion itself often. Despite the fact that infinity is in its nature incomprehensible to our limited intellects, we may nevertheless arrive at a number of profound conclusions regarding that nature—indeed, from my experience, familiarity with a subject is made all the more desirable when it proudly evades our grasp.

Salv. Very well. It will be necessary then, Sagredo, for the sake of Literasto and for ourselves, that we review our previous demonstrations, from which resulted those strange conclusions you refer to.

Sagr. That should not present a great problem, since they have been fresh in my mind for some time now. And I am glad to be of service to your friend Literasto, who will surely provide for us a critical eye where it is most needed.

To begin, I recall Salviati first concluding that an infinite number of unquantifiable parts may actually constitute a magnitude. In other words, that is to say that an infinite number of infinitely small, indivisible points might actually form a line.

To quote Salviati himself, if I remember correctly, "not only two indivisibles, but ten, or a hundred, or a thousand do not compose a divisible and quantifiable magnitude; yet infinitely many may do so."[4] Moreover, it was seen that this magnitude, being composed of infinitely many unquantifiable parts, might be of any given extension.

---

[4] Ibid., 39.

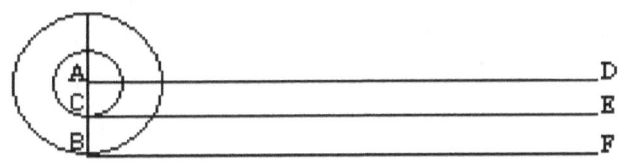

To show this he examined two concentric circles and, supposing the larger to have been rolled along a straight line BF equal to its circumference, it was understood that the smaller circle, in rotating with the larger, also describes in one full rotation a straight line CE equal to the circumference of the larger.[5] Since both circles, however, are composed of infinitely many points, one can see that the points of the smaller circle, that is, its circumference, are in this case made to equal in length the circumference of the greater circle.[6] Therefore the points that make up the circumference of the smaller circle might constitute a line equal to any given length.

Lite. Before we go any further, allow me to bring up an objection to what has already been set forth. From the start, it seems that you are taking as granted that an

---

[5] Ibid., 31.

[6] The theory of "void points," which is not crucial to the mathematical demonstration presented, is omitted here for the sake of simplicity.

infinitude of points actually exists in a given segment of a line, and that the line is indeed composed of these points. Whereas I recognize the continuous and the discrete as two distinct species of quantity, this hypothesis seems to hold that they are really convertible—that magnitude is nothing more than multitude. But how can any multitude of infinitesimals—I mean those points that really have no extension[7]—possibly compose a magnitude? Is this not flatly illogical?

Yielding to the possibility, at least in theory, of rolling a perfect circle along a perfectly straight line, this construction still does not suffice to prove that infinitely many points might construct any given straight line, since it is far from clear to me that such points can assemble anything at all. I would not be so rash, now, as to assert that the number of points within a line are finite; rather I follow Aristotle, who writes, "A quantity is infinite if it is such that we can always take a part outside what has been already taken."[8]

---

[7] Cf. Euclid, *Elements*, trans. Sir Thomas L. Heath (Encyclopedia Britannica, 1996), 1. (Bk. 1, Def. 1)
[8] Aristotle, *Physics*, trans. R.P. Hardie and R. K. Gaye, ed. Richard McKeon (New York: Random House, 1941), 3.6.207a8.

In this way the divisions of a line are truly infinite, but those points are not actually present until the line is divided. Since the divisions of a line are never exhausted, those infinitely many points can never be present all at once, even as the entire day cannot be present to us in its entirety. And in fact I see no other way to approach the matter than this.

Sagr. I do not wish to dwell on pedantic arguments that have no real bearing on the extraordinary conclusions that you draw, Salviati, but it appears to me that Literasto has been overly liberal in granting you the possibility of rolling a circle along a line. For it strikes me that since locomotion cannot exist apart from physical things, the circle must be rolling along this line in much the same way the wheels of a carriage roll down our Venetian streets. But when I try to imagine a standing circle starting its revolution, I see that it must move from the point on which it currently stands to some next point on its circumference. Clearly, however, if the points of the circumference are infinitely many, there can really be no next point onto which the circle can roll, and so motion in this case seems impossible.

Salv. I am glad that you have voiced your objection, Sagredo, rather than suppress doubts in the back of your mind; for if motion in this machine would be impossible then the conclusions drawn from it must be dubious at best. Let me remind you, however, that Zeno applied this same objection to motion in general, inquiring how it is that anything might move from one point to another, since it must traverse an infinite number of points in doing so.[9]

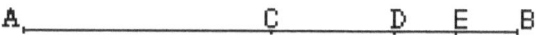

In moving from point A to point B, for example, an object must first pass through the midpoint of that length, C; and in moving from C to B it must again pass through the midpoint of that length, D; and, continuous quantity being always divisible, this process will continue ad infinitum, the object always moving toward some midpoint between its own position and B. Thus he concludes that the object must always approach and never attain its final destination.

---

[9] Cf. ibid., 6.9.239b12.

Lite. I see that this so-called proof may be very puzzling for those who have no familiarity with ancient wisdom on the matter, even motivating some to desperately declare that motion is not really continuous at all. For each additional length that is taken is "always finite, but always different,"[10] and if anywhere the infinite manifests itself it is in this case. It must be understood, however, that the moveable never rests at any of those points in between, but instead passes through them continually.

Having spent no time at any midpoint, the path on which the object travels is not actually divided at all, but those infinitely many divisions remain in potency. This is how Aristotle reveals the paradox to be a mere phantasm, for "motion if it is to be continuous must relate to what is continuous."[11] For this reason I did not object to the possibility of rolling a circle, though my mind is strained to imagine your example in its peculiarities.

Salv. Not begrudgingly do I concede to you this explanation, good friend, since I know that you have much experience in

---

[10] Ibid., 3.6.206a29.
[11] Ibid., 8.8.263a27.

putting such persistent beasts to rest. It is important to note that where the motion of the moveable is divisible into its infinitely many parts, so too must be the time during which it moves. For the paradox of Zeno did not set itself against the mere possibility of an object reaching its destination, but it was rather against the idea of that object reaching its destination in some finite amount of time. Yet it must be recalled that time, like distance, is a continuous quantity, and therefore no less divisible than the straight line.

If one would treat the motion of an object as traveling through its infinitely many points, he cannot then ignore the infinitely many moments during which it also travels. I say, therefore, that there is no reasonable objection to the crossing of a moveable over infinitely many points, actual or potential, provided that one also recognize this process as taking place during an infinitude of times. But to the question of whether an object can be at a place in which it spends no amount of time, I would fairly reply that it can—for a moment.

Now if the indivisibles of time were to correspond to the indivisibles of motion, point to point, it may at first seem that in a continuous motion only one uniform speed is really possible. Here, however, rather than finite quantities, we are dealing with an infinite number of times and an infinite number of distances. Because such infinitely many unquantifiable parts might actually compose any given magnitude, as shown by the demonstration that Sagredo put forth, it follows that any given velocity may thus be obtained as produced from some finite distance and some finite time. In this way we may offer some account for the various speeds with which we see things move all around us.

Here I must stop and ask Sagredo whether he may now put away his anxiety about our experimentation with rolling circles, or if instead we have only stirred up some other unseen troubles in the course of our explanations.

Sagr. By both accounts I must admit that my own objection has been quite deftly disarmed, and yet I can hardly claim satisfaction. I am instead beginning to wonder whether I am getting, as they say, 'more than I bargained for,' and whether perhaps a clear understanding of this theory of infinites is simply beyond my reach. As I cannot object to my having walked right up to your doorstep tonight, Salviati, I will not deny that these circles can roll; but nor do I understand just how this happens. Already my head is spinning, and I fear that we still have much work to do.

Salv. The truth of the matter is without doubt covered in many obscurities. Nevertheless, a subject of such magnificence surely cannot succeed in hiding itself entirely from human eyes, and I am confident that, with due perseverance, we will at least be rewarded with some small glimpse of its character before the night is done. And do not be surprised if what now appears to be too complex for the human mind to consider eventually reveals something much simpler than expected. For nature loves to adorn herself in a great many ornaments in appearing before man, and yet if we follow her home at the end of the day it may happen that we catch sight of her beauty clothed in nothing more than a simple nightgown.

But allow me now to address those first objections raised by Literasto, who has shrewdly chosen to ignore the squabbles surrounding the theory, lunging instead straight at the heart of the matter. And in doing this I hope to lay out more clearly my own thoughts on this subject. You, Literasto, suggested that it is absurd to think that a line is composed of infinitely many points, and were wary to conclude that discrete and continuous quantity are really of the same constitution. Tell me, then, what kind of thing you consider a circle to be.

Lite. I would refer such questions primarily to the writings of Euclid, whose definitions I deem especially worthy of memorization. He defines the circle as "a plane figure contained by one line such that all the straight lines falling upon it from one point among those lying within the figure are equal to one another."[12]

Salv. Very well. The circle is a figure then, and it is contained by one line. In this case the line is said to be curved. But what is a line?

Lite. "A line is a breadthless length."[13]

Salv. Good. Now, the straight line certainly exists in some way, though it may not be clearly manifested in physical things. But even those intangible lines existing in our head are not entirely breadthless, and neither,

---

[12] Euclid, *Elements*, 1. (Bk. 1, Def. 15)
[13] Ibid. (Bk. 1, Def. 2)

perhaps, will they be perfectly straight. If we cannot properly imagine a straight line, how then do we know that such a thing might exist?

Lite. I know that you agree with me, dear Salviati, when I say that only a fool would deny the straight line a theoretical existence. Nevertheless, I can say that what I am trying to imagine is "a line which lies evenly with the points on itself,"[14] and as I can find no contradiction in the terms of this definition, I therefore have no reason to doubt the existence of that which it defines.

Salv. The straight line, then, 'lies evenly' with the points on itself. Yet to lie evenly, it seems, is nothing other than to be straight. Be that as it may, we will avoid foolishness by saying simply that we identify the straight line by our concept of straightness. But if we know straightness from the start, then length must be known as well, because straightness presupposes some length. Evidently, then, familiarity with the straight line is inherent in us. But what can we say of the curved line? What is its definition, and how are we to identify it?

---

[14] Ibid. (Bk. 1, Def. 4)

Lite. Although Euclid has neglected to define the curved line, we may infer it to be a line that does not lie evenly with the points on itself. We must not, however, include the 'bent line' in our definition of the curved, for in that former case there are really two lines present, while in the latter only a single line is assumed.

Salv. Let us mull over the distinction between bent and curved lines with careful scrutiny. For if there is not but one assignable length between two given points—if rather there are distances between those points as numerous as the curved lines joining them—then we surely ought to pity the daily labor of our poor carpenters, since in that case there will be as many lengths as you please answering to any question of measurement. For their sake and ours, let any use of the word 'length' restrict it to the straight alone, and predicate it only of that which has straightness in it.

The bent line is assigned a length equal to the sum of its parts, and so the perimeter of a hundred- or a million-sided polygon. But tell me, by what argument is the circumference of a circle admitted to have any length at all?

Sagr. It seems to me that all those curved lengths you refer to only exist in potency when the line taken between the points is straight, but why can we not instead begin with a curved line? In that case it is the straight line, I think, which is found to be immeasurable; and so the proof works both ways. For a circumference presents itself to me as necessarily having some length, inasmuch as it is the boundary of a circle, and even if we cannot compare the curved to the straight, I see no reason to sacrifice one concept to the other.

Salv. Such language, my friend, is the symptom of a contagious malady, of which our Academician[15] once found himself a victim. Understand first that a curved line cannot be prior to its length; rather, having no measurable breadth whatsoever, the line is nothing more than length itself. Having accepted the definition provided us by Euclid, we then find ourselves at a complete loss to show that our supposed 'curved line' is either breadthless *or* a length. Allow me to explain.

First, we recognize that a thing has both length and breadth when it has more than one

---

[15] Galileo.

extension in space. The circumference of a circle, comprehended as a whole, clearly has multiple extensions. If we treat of the bent line in its entirety we must come to the same conclusion—that it requires both length and breadth. Yet we may call the bent line breadthless because we know that its parts can be taken in such a way that none of them, considered independently, have any measurable width at all. This cannot be in the case of the circumference because any finite part of it, no matter how small, still requires two dimensions for its existence. It is only at the points of the circumference that we find true breadthlessness.

Second, as locomotion is impossible without movement in a certain direction, so is length nonexistent without extension in a certain direction. As it is, extension in a direction exists nowhere along the circumference of a circle. Again, if length does exist in some way along that circumference, it is only through the points. Why then should it surprise us that all attempts to measure curved lines by straight, or even one curved line by another, are utterly futile?

Sagr. You are a stingy man, Salviati, to have kept such a magnificent gem hidden away from your friends for so long! The proof has been succinctly delivered, but I doubt that its meaning will be so easily comprehended.

Lite. I think that I understand now the root of the problem in trying to determine the measure of a curved line relative to the straight. I am, however, most hesitant to agree when you say that length can only be said of what is straight, because what follows from this as a conclusion can fairly be called an abomination to science. As things now stand, the circle is the basis of all geometry. Yet if its circumference is everywhere devoid of straightness, and thus length as well, then the circle cannot be bound by a line; and if not enclosed by a continuous line, then it is not properly a geometrical figure at all. In that case the circle would be bound by nothing but points!

Salv. It would be disastrous indeed if such arguments as these compelled us to unravel the fabric of sciences woven over so many ages with such delicate care. Yet we are freed from this obligation in acknowledging that both the straight line and the circumference of the circle can be treated of in their composition from infinitely many points, which points perhaps contain something of straightness in them.

Sagr. I have not forgotten your earlier account of circles as "polygons of infinitely many sides,"[16] and I see that the justification for this description is found under the light of these new considerations.

Lite. Hold the horses for a minute, and tell me plainly whether you believe circles to be bound by a line or by infinitely many points. I have followed your arguments, which admittedly are somewhat unsettling to the mind; but just a short while ago you did agree that no given number of points could ever constitute a line, and so I am utterly confused now to hear each and every point of the circle being referred to as a 'side.'

Salv. As to that question of whether the circle is bound by one side or by infinitely many, I think that to presume these two answers to be mutually exclusive would be a great error. Astonishing as it may sound, we ought rather to reply that both assertions are equally correct.

---

[16] Galileo, *Two New Sciences*, 33.

Lite. I have come to expect such responses from you, Salviati, and though at the moment this idea sounds absurd to me, yet I eagerly await your explanation.

Sagr. Allow me here, Salviati, to employ the same reasoning for Literasto that you made use of earlier today to occupy our own Simplicio, that you may save your breath for its further explanation afterward.

Salv. In that case I will relax and enjoy the exchange; only I hope not to disappoint when all is said and done.

Sagr. In my experience such has never been the case. But do not hesitate to interrupt should I omit anything essential.

Now, that there is no last among numbers, or, that there is no number greater than all others, is a universal belief among men that is so far from controversy as to have been disputed, to the best of my knowledge, by no significant authority in history. And if one were to challenge this belief, we should press him to reveal to us that highest number, to see whether we truly cannot add one more unit to it and thereby end up with a greater figure still; but our attention could only be paid in mockery, for it is perfectly clear that any number given can be added onto quite easily.

Lite. Very true.

Sagr. Well then. Let us now consider those integers of the number series called 'squares.' A square number, produced by the multiplication of a number into itself, is struck upon every so often as one counts along the number series. Four is a square number having two as its root number; nine is a square having three as its root; sixteen is the square of four, and so on. A square number is observed at certain intervals of the number series, and the many numbers that are passed over in this interval are not square numbers at all: five, six, seven, and eight, none of which are produced by the multiplication of a number into itself, are all included in the interval between the square numbers four and nine.

Moreover, as the numbers are counted further and further upward, the square numbers become less and less frequent, and the intervals of nonsquares between them greater and greater. Within the first hundred numbers there are ten squares, which constitute one-tenth of these hundred; up to ten thousand, only one hundred are squares, constituting one-hundredth of those counted; and for the first million numbers, only one thousand are squares, and these squares constitute one-thousandth of the total numbers, so that the ratio of square numbers to all numbers decreases as the counted numbers increase.

Lite. I follow.

Sagr. On the one hand, then, it appears that the list of all counted numbers must be much *longer* than that of the square numbers alone, since every square number must surely be included on the list of all numbers, and yet there are plenty of numbers that will never appear on the list of squares. On the other hand, every given number must necessarily have a square corresponding to it; that is, every number is a root that can be multiplied into itself to produce its square. So it also seems that there cannot be *fewer* squares than there are numbers. Each side pleads a compelling case. What is your judgment, good Literasto?

Lite. I do not see that arbitration is necessary here, for the impossibility of ever collecting such a list of 'all numbers' makes the entire trial ludicrous. To be sure, both lists can be continued as long as you like, in which case the list of all numbers will always be the greater, as the whole is always greater than the part. The strange belief that these numbers exist on their own, wholly independent from the things that they number, is in my opinion the cause of this and other needless anxieties. But only bring me those lists, dear Sagredo, one of all numbers and another of all squares, and I will tell you which is the greater.

Sagr. It is admittedly beyond my abilities to provide the requested catalogues, but has not Salviati just now presented us with a sufficient example of an actual infinite? For in the circle we have not a quantity that is always finite and always capable of being added to, but rather an infinite multitude that is fully present in its entirety! So if you will allow me to represent the list of squares by the points on the circumference of one circle, and again the list of all numbers by the points of another, larger circle, then you must admit that you still owe me an answer to that perplexing question.

Or, if a clearer case is desired, allow me to recapitulate the brilliant argument that Salviati previously brought against our Simplicio—for Simplicio too was of the mind that the infinitely many parts of a straight line were contained in it only potentially.

In dividing such a line, Salviati begged whether, "the actual division of such parts having been made, that original whole has increased, diminished, or remains still of the same magnitude?"[17] To which Simplicio quite reasonably replied, "It neither increases nor diminishes."[18] And so it was concluded that the quantified parts of the continuum, whether potentially or actually there, in no case make its quantity greater or less. But if quantified parts are actually contained in that line, and they are infinitely many, then they make that line to be of infinite magnitude; whence "infinitely many quantified parts cannot be contained even potentially except in an infinite magnitude."[19] For what exists in theory is not altered by the passage of time, but is as present today as it will be tomorrow. If the parts of the straight line are at all infinite, this must be so by virtue of its indivisible, or unquantifiable, components.

Lite. Forgive me if I am once again thrown into doubt about this matter, when just

---

[17] Ibid., 43.
[18] Ibid.
[19] Ibid.

moments ago I led you to believe that your patient labor had found a new convert. But what am I to think?

You hold that these infinitely many numbers and squares are present in their entirety. Then clearly there *cannot* be more numbers than squares, for each number has exactly one corresponding square; and yet there *must* be more numbers than squares, lest the whole be not greater than the part![20] Such an ugly conclusion must be traced back to its parent premises, for our logic is sound. An 'infinitely many,' present in its entirety, cannot be. Or have you a *deus ex machina* prepared for the situation?

Sagr. The only reasonable solution to be found for this dilemma was in concluding that "the multitude of squares is not less than that of all numbers, nor is the latter greater than the former,"[21] and it was finally decided that "the attributes of equal, greater, and less have no place in infinite, but only in bounded quantities."[22]

Lite. We ought not to rely on abstract expressions to whisk us out of a difficult

---

[20] Cf. Euclid, *Elements,* 2. (Common Notion 5 [8])

[21] Galileo, *Two New Sciences*, 41.

[22] Ibid.

situation. Although I feel a certain attraction to it, yet I fail to see how this explanation avoids a conflict with Euclid—for if the multitude of all numbers is in fact not greater than that of the squares, then neither is the whole in this case greater than the part.

Sagr. Our present state of affairs recalls to my mind an interesting diagram used by Salviati to illustrate just how great a difference exists between the finite and the infinite, and with what care we ought to treat of infinite quantities.

Let a straight line AB be taken, and let this line be divided at point C into unequal parts AC and CB. Let there also be described many pairs of straight lines: in each of these pairs let one line touch point A and the other point B, and let also the lines of each pair touch each other, and the ratio of AC to CB be preserved in each couple; for example, AI and IB, AG and GB, and AF and FB. Our generous host demonstrated that the meeting points of each of these complementary pairs all lie on the circumference of one circle, CIGF.[23]

[23] Ibid., 51–3.

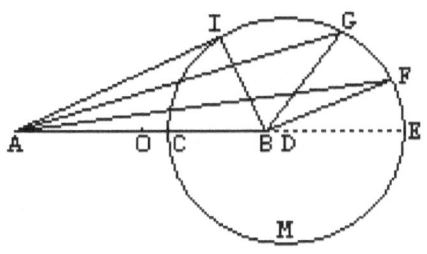

Now, the circle described in this way will be ever greater as the point C is taken closer to the midpoint O of the original line, and ever smaller as C is taken closer to point B. And if we take point C *at* point B, the resulting circle will be infinitesimal—that is, nothing more than a point.

We ought to expect then, conversely, by taking point C *at* O, and following the same rules as set out before, that we will thereby obtain a circle of infinite size—and what is actually traced out by the meeting points of each of those pairs is a straight line, perpendicular to line AB at point O, and extending in both directions without end!

As Salviati put it, the circle, in passing from a finite to an infinite or infinitesimal size, "changes its being so completely as to lose its existence and its possibility of being [a circle]."[24]

---

[24] Ibid., 47.

Salv. It was wise of you, Sagredo, to return at this point to a consideration of the kind of disparity that exists between the finite and the infinite; for the difference in passing from one to the other seems to involve, rather than a change in the quantity of a thing, a change in its very nature. Whereas a change in quantity modifies *how much* of a thing we have, a change from finitude to infinitude seems instead to modify *what* we have.

Up to this point we have spoken of an 'infinitely many,' debating whether or not such a multitude might exist without perhaps understanding what that existence would imply. The problem of the whole being always greater than the part makes for a puzzling situation indeed; but we, like the practiced shipbuilders at our magnificent harbor, must finally put our subject to the test, to judge under what circumstances it will stay afloat and under which it will sink. For no matter how great are the numbers that we choose, it is always possible to compare one multitude to another by virtue of some unit. But when our multitudes pass into infinity—just as when our circle became infinite—the question must be raised of whether the resulting quantity is able to retain its nature as a multitude.

Sagr. I feel that we are approaching the culmination of our investigation into the nature of infinity, for we have at last turned to the point that has vexed me this entire evening. There were certain particularly insightful observations made by Salviati in our previous discussion pertaining to the unity of the infinite. I took care at that time to copy his language down word for word, and as I could not hope to deliver it more eloquently than did our host, here it is in its original form:

In our discussion a little while ago, we concluded that in the infinite number, there must be as many squares or cubes as all the numbers, because both [squares and cubes] are as numerous as their roots, and all numbers are roots. Next we saw that the larger the numbers taken, the scarcer became the squares to be found among them, and still rarer the cubes. Hence it is manifest that to the extent that we go to greater numbers, by that much and more do we depart from the infinite number. From this it follows that turning back (since our direction took us always farther from our desired goal), if any number may be called infinite, it is unity. And truly, in unity are those conditions and necessary requisites of the infinite number. I refer to those [conditions] of containing in itself as many squares as cubes, and as many as all the numbers [contained].[25]

You will note that we work here only with those numbers that truly exist, having adopted the Euclidean definition of number as "a multitude composed of units,"[26] where a unit is "that by virtue of which each of the things that exist is called one."[27]

The business has in it no room for doubt, because unity is a square, and a cube, and a fourth power, and all the other powers. There is no essential property belonging to squares, cubes, and so on that does not belong to [the number] one. For instance, a property of two square numbers is that of having between them a number [that is their]

[25] Ibid., 45.
[26] Euclid, *Elements*, 127. (Bk. 7, Def. 2)
[27] Ibid. (Bk. 7, Def. 1)

mean proportional. Take as one extreme any square number, and as the other, unity; there will always be found a numerical mean proportional; thus let the two square numbers be 9 and 4; between 9 and 1 the mean proportional is 3, and between 4 and 1 it is 2; between the two squares 9 and 4 we find 6, the middle [term in geometric proportion]. A property of cubes is that between them there are necessarily two mean proportionals; given 8 and 27, between them lie the [geometric] means 12 and 18; between 1 and 8 are 2 and 4; and between 1 and 27 are 3 and 9. Thus we conclude that there is no infinite number other than unity.[28]

Lite. Well, this is a curious change in the weather! It seems that we have come around from positing an infinite multitude of points in any given line to concluding that there is really as much of a possibility of having an 'infinitely many' as there is of having a 'square circle.' But I will not put words into your devious mouth, Salviati. By all means, let us have out with the rest of the theory.

[28] Galileo, *Two New Sciences*, 45.

Salv. It is true that, taken in a strict sense, the 'number' one is not at all a number—on the contrary, *one* is quite the opposite of number. But infinity has from antiquity been thought to belong to the genus of quantity.[29] And to me this seems proper, for *it* is that by virtue of which each of the things that exist is called one. Where quantity deals with the measured or measurable, infinity seems to be a principle of measurement.

And I think that a careful inspection into the meaning of the word 'infinite,' that is, 'without limit,' bares the same results. Now what is one in quantity may initially seem to be so in *respect* of its limits. But in order to consider a complex thing as one, we must *ignore* all limits that it has within itself, that is, what separates one part of it from another; for insofar as we acknowledge those divisions we understand the thing not as one, but as many. If we are, for example, to consider a divided line segment as one thing, we must disregard the divisions that it has within and pay heed only to its outermost boundaries; and in this act we treat that segment as continuous.

---

[29] Aristotle, *Physics*, 1.2.185a33.

But that we may avoid, as they say, 'getting ahead of ourselves,' allow me now to address the problem of the whole being greater than the part. The apparent conflict lies in the idea that one multitude is at the same time both equal and unequal to the other. But I believe that this difficulty is in fact resolved by that very movement from finite to infinite quantities, for, as we have said, the infinite is not a multitude, but a unity.

It is very significant that when we attempt to compare infinites by virtue of their multitudes we find this to be impossible. We speak of one unit being greater than another, but not, of course, in virtue of multitude. Is a magnitude greater, less, or equal to a number? Likewise the infinite in numbers is not greater, less, or equal to the finite, as Sagredo pointed out in our discussion with Simplicio.[30] But two infinites considered as magnitudes very well may have that relation of part to whole. And in this way our problems are resolved.

You have patiently followed my account up to this point, Literasto, but I perceive that some thought agitates you now. I implore you to articulate it for us, if you are able.

---

[30] Galileo, *Two New Sciences*, 41–2.

Lite. At the moment I am pondering no longer a single continuous line, but rather a multitude of broken line segments stretching off in either direction as far as the eye can see. Let us grant that these are actually infinite in number, as we are now permitted to do. If I understand you correctly then these segments ought to be unified in some way…

Salv. That is most certainly true, for if it were not so then your magnitudes could not be 'infinitely many' to begin with. We have shown that the infinite in numbers is found only in the number one, and in dealing with many magnitudes that go on without end, with spaces interspersed between them, I think that you will find two ways in which these innumerables can be considered: either as individually numbered, by which we focus on some finite number of segments at a time, or as a whole, by which we attempt to wrap our mind around all of them at once. But clearly any consideration of these segments in their multitude will not do justice to their infinitude.

We must then treat of all of your supposed magnitudes simultaneously if we hold them to be actually infinite. But under my reasoning there is only one way in which those infinitely many line segments can all be present together: through the principle that determines them. Indeed, it seems to me that the true unity of any infinite quantity is really to be found first and foremost in the principle that originates that quantity.[31]

Now when the quantity under consideration is a single continuous magnitude, such as a line segment, its unity is boldly manifest to us. But the situation is made more difficult in the case that Literasto has provided, for we cannot conceive of many separate magnitudes as being in any way unified except by this aspect: that they follow a certain pattern. And so we may find satisfaction in concluding that the principle of this pattern is the selfsame principle by which those magnitudes are infinite—that is, the principle of their unity.

Lite. You must forgive me my stubbornness on this one point, Salviati, but suppose that there simply is no pattern to be found among these magnitudes. Where will the unity lie in that case?

Salv. It is my belief that there can be no infinite where no determining principle is

---

[31] Cf. ibid., 181–2.

assumed. For as Sagredo has pointed out, that which exists in mathematics is not bound by the shackles of time; so if those infinitely many magnitudes are present all together in a certain order, this can only be through that which orders them. To my mind it is as a fountainhead dividing one stream of water into many, according to its special design. In this way does the originating principle determine which points go where, whether these are continuous, and at what intervals they occur.

Sagr. I think that I am now beginning to understand just what Salviati means when he insists that unity and infinity are not really distinct ideas. The question of how the circle is truly one geometric figure has lingered in my mind since our examination of its 'infinitely many' points. But by this new reckoning I believe we have the answer: for if every infinite quantity has its own determinate principle, then the circumference of the circle must surely have its origin in the very definition of the figure, that is, in the notion that the points of its circumference are all equidistant from some one point. And this is ultimately the reason that the circumference is one, and why a denial of that unity is so repugnant to our minds.

Salv. Your genius, Sagredo, surely lies in the patient and diligent reflection that you invest in every idea presented to you. But the extraordinary nature of the circle is not limited to the infinitude of its points alone, for the straight line has this character as well. The circle seems to be unified in yet another way: that in its boundary are contained not only 'infinitely many' points, but 'infinitely many' angles as well! For at absolutely no place on a circumference is straightness observed, and neither are there any irregularities in its curve. And so, contrary perhaps to what previously seemed the case, we now have reason to believe that within a plane there indeed exists no shape *more* unified than the circle—and this, the simplicity of its principle, I judge to be the true sublimity of that figure.

Lite. Congratulations are in store for you, Salviati, for I have found your recent sermons to be as delightful to the intellect as they are to the ears. And yet I see that they will require more consideration than can be given in the short time left to us tonight.

Sagr. My friends, I have no doubt that there are many other profound insights to be had regarding things infinite, especially under the guidance of our new theory; but I fear that we must end our discourse for now if we seek sufficient recuperation during the night for the tasks we have set before us on the morrow.

Salv. Only stave off that drowsiness for a minute longer, good Sagredo, as I cannot allow you to depart tonight without having released this last thought from the forefront of my conscience. We have dealt thus far with two kinds of infinite quantities: that which is ultimately contained but is not within itself divided, and that which does have limits within but has no outer boundaries. It must here be inquired whether it is possible for something to be *absolutely* infinite, that is, having limits neither within nor without of itself. And it seems to me that if something of this nature does indeed exist, then neither can anything lie outside of it, nor will anything divide it within; and, most interestingly, it would seem to be a principle of all things quantifiable.

Lite. The transcendent nature of our subject calls to mind the teachings of great theologians who labored hard in their own field for a taste of the divine. And these propositions they held true of God: that He is both infinite and "supremely one,"[32] and also that He is utterly simple; that He exists everywhere at all times, such that all things are present to Him simultaneously in their entirety, and also that He is not in place, that He is outside of time.

Salv. Still other thoughts occur to me at the moment, the significance of which seems to overshadow much of what we have said thus far; but it would seem a kind of sacrilege to drag these ideas down from their proper place of contemplation in the heavens in an attempt to set before us in speech what cannot be properly spoken.

Alas, we have discussed these things late into the evening, and, as Sagredo has prudently pointed out, we must now rest so as not to disappoint Simplicio in the morning.

Sagr. I will certainly look forward to that meeting in the meantime.

The Evening of the First Day Ends

---

[32] Saint Thomas Aquinas, *Summa Theologiae*, trans. Fathers of the English Dominican Province (Encyclopedia Britannica, 1952), 1.11.4.